LET'S GO TO CHURCH

WRITTEN AND ILLUSTRATED
BY
Bessie Dean

INTERNATIONAL STANDARD BOOK NUMBER
0-88290-046-3

Second Printing
June, 1978

Printed in the
United States of America
by
**Horizon Publishers
& Distributors
P.O. Box 490
50 South 500 West
Bountiful, Utah 84010**

Contents

About the Author

Bessie Ann Dean was born and raised in Salt Lake City, Utah. Her love for children has been a guiding influence in the many positions she has filled in the auxiliary organizations of The Church of Jesus Christ of Latter-day Saints.

Sister Dean has been a Y.W.M.I.A. Stake Board member, a Counselor in Primary and Relief Society, a Primary President and a Sunday School Teacher. Because of her musical abilities she has served as a Junior Sunday School Chorister and has also been Chorister for the Sunday School, Relief Society and Primary. For the past six years, she has served as the Junior Primary Chorister.

Her husband, Gary, is a Bishop's Counselor. They have three children: Scott, Bob and LeAnn.

Sister Dean has devoted much of her time to the teaching of young LDS members. The ability to tell stories that teach an important message, and which will hold the interest of pre-school children, is a rare gift. In this book Sister Dean combines that talent with her artistic skills—she is both the author and illustrator. LET'S GO TO CHURCH is one of a special series of books for young Latter-day Saint children which is to be released by Horizon Publishers.

Sunday
Is Special

Michelle just got out of Sunday School and could hardly wait to get home. Her teacher had given a lesson on keeping the Sabbath Day holy, and she wanted to tell her family about it. She had known that Sunday was the Lord's day, but had wondered just what she was supposed to do, and now she knew.

The first thing she did when she got home was help her mother get dinner on the table. She had helped her mother prepare the food on Saturday so they wouldn't have to do any extra work on Sunday.

Michelle was so excited! She had so many neat things to tell her family.

After they said the blessing, Michelle asked her brother, Mike, if he knew what they were supposed to do to keep the Sabbath Day holy. He said that we are supposed to go to Church and partake of the Sacrament. He was right! That is the most important thing to do on Sunday. Then Michelle told them the special things she had learned.

Besides attending all our Church meetings, we can study the Scriptures. Michelle liked this idea because it reminded her of Family Night when all her family got together. She liked to hear Bible Stories and learn more about Jesus.

9

The next thing that can be done on Sunday is visit our relatives. This sounded like fun also. Michelle loved to go to her Grandma's house and visit. It seemed like she always had a cookie and a big hug for Michelle and her brother.

Michelle could remember one more thing to do to keep the Sabbath Day holy. That was to visit the sick.

10

After Church, Michelle and her family decided to visit sick, Sister Jones. It made her happy to have someone visit her. Michelle thought how lonely she must be just having to stay in the house all the time.

Mike and Michelle both had a happy feeling inside when they saw how much it helped Sister Jones. They were glad they had cheered her up.

Later that evening, Michelle's dad called all the family together to study the scriptures.

Michelle liked the stories her dad read. She and Mike both learned a lot about the Gospel.

Before she knew it, it was time to get ready for bed. This had really been a special day. Some Sundays Michelle had been unhappy because she had wanted to go outside and play ball, but her dad had told her he didn't think that was a good thing to do on Sunday. Now that she knew just what to do to keep the Sabbath Day holy, she would be able to have a special Sunday each week.

When she kneeled down to say her prayer before getting into bed, Michelle thanked Heavenly Father for her family and the special day they had had together. Now she could hardly wait until next Sunday.

Herman Goes
To Sunday School

Herman was a little green frog who lived in a field by a Church. He was a happy little frog and he and his friends really had a lot of fun in the field. There was a pond to splash around in, rocks to jump over, and bushes to play hide and seek behind. Herman was usually very happy except for one thing.

Every Sunday the boys and girls came out of the Church with such big smiles on their faces that it made Herman curious to know what happened in the Church to make the children so happy. He was afraid he might be missing something fun. He just had to get inside that Church and see what they did!

17

One day Brent and his friend, Kenny, were playing in the field. Herman hid behind a bush and listened to what they were saying. They were talking about the things they had been doing in Sunday School. Herman really listened closely when he found out what they were talking about.

Herman stayed awake almost all night to figure out a way to go to Church. Finally he got an idea! He would hide behind a rock until Brent came through the field on his way to Sunday School. Then he would jump into the cuff of his pants and go to Sunday School too. Herman was so excited he could hardly wait for Sunday to come.

Finally Sunday came, and sure enough, here came Kenny and Brent, all dressed up.

As soon as they got by the rock, Herman made a great big leap—and missed! He landed right on the ground! He surely hadn't expected that to happen. That was not the way he planned it at all! He picked himself up, brushed the dust off, and wondered how he would ever get to Church now.

He looked up to see where the boys were and was he in luck! They had stopped to look at a bug. Herman hurried along and tried his special leap again.

This time it worked. He landed right in Brent's cuff. Now he was on his way to Church!

Just before the boys got to the door, something happened. Brent looked down and saw Herman in his cuff. He took him out and put him on the ground. He told him that frogs can't go to Church—only people can.

Herman felt so sad that he was just about ready to cry, but then he got an idea.

He decided he would sit on the ledge by an open window and then he could peek in and see what the children were doing in Church.

As soon as the boys got inside the doors they quit yelling and pushing each other. Herman wondered what they were doing, but then he heard someone talking about being reverent in Heavenly Father's house. He decided that must be what being reverent meant.

He soon learned more about rever-ence as the boys walked quietly into a room and sat down.

Someone was playing some soft, pretty music and all the boys and girls he could see were sitting in their chairs with their arms folded.

Herman had a good feeling inside and decided he liked to be reverent.

Pretty soon the children sang a song and then a little girl said a prayer. A nice lady talked to the children about the Sacrament and how they should think about Jesus and the things He had done for them. During the Sacrament the children tried very hard to be extra reverent.

After the Sacrament was over, it was time to sing a song. Herman thought this was really fun and even tried to sing. All that came out was a funny, croaky noise. A boy named Erik turned around to see where the noise was coming from. Herman tried to hide his face so no one would see him. He decided he would have to be more careful and not get so excited and sing so loud. He really liked to sing the songs.

Herman enjoyed the rest of Sunday School. A boy gave a talk about being kind to one another and a girl told a story about a little boy who paid his tithing because he knew it was the right thing to do.

All too soon, Sunday School was over and the boys and girls were in the field again on their way home. Herman hopped down off the window ledge and went back to his home. He felt happy inside because he had learned so many good things about being kind and helpful to one another. He knew now why the children looked so happy when they came out of Church. They were learning the right things to do so they could live with their **Heavenly Father** again.

Herman was sitting on his favorite rock with a great big smile on his face when his friend Mort, the toad, came along and looked at him. He said, "Herman, why do you look so happy? What have you been doing?"

Well, Herman could hardly wait to tell him that he had found out about Sunday School.

Bob Learns
To Be Reverent

Bob **sat straight** up in bed! He had just had the strangest dream. He dreamed he had gone to Church, but something was wrong. Everything seemed so different.

When he went into the chapel the organ wasn't playing and the people looked very unhappy.

He saw his teacher, but she wasn't smiling as usual. She had a very sad look on her face. Bob asked her what was wrong and she told him that some children had forgotten that they were in Heavenly Father's house and were running around instead of being reverent. The teacher knew that they had only forgotten for a little while, but it made her feel sad because she knows that the Church is a very special place.

Just then Bob woke up.

Later in the day he saw his friend Laura, and told her all about his dream. They thought about it for a little while and then got a really good idea. They decided to try and be more reverent in Church and help their friends do the same.

When they got inside the Church building, they walked very quietly in the halls. One little boy started to run, but when he saw Bob and Laura he stopped running and walked reverently instead. Laura picked a piece of paper up off the floor and put it in the wastebasket so the Church would be neat and clean.

While the speaker was talking they listened quietly instead of talking to each other. They were surprised at how much they learned.

When it was time to sing, Bob picked the hymnbook up very carefully so it wouldn't tear and sang the song. When the song was finished, he put the book back in its rack very carefully.

Bob noticed a funny thing. When he was being reverent, some of the other boys and girls followed his example and everyone was more reverent.

When Church was over and Bob was walking out to the car, he turned around and saw his teacher. She looked a lot different than she had in his dream. Now she had a big smile on her face. She told Bob how proud she was of him for being reverent in Church. Bob had a good feeling inside.

Bob was glad he had had his strange dream because it made him think more about being reverent in Heavenly Father's House.

Jodi Takes The Sacrament

It was Sunday morning. Jodi woke up early so she would have plenty of time to get ready for Sunday School. She really liked Sundays because she could go to Church and learn more about Jesus and the Gospel. It was really a special day for her.

She hurried and put on her best dress so she would look nice to go to the Lord's house.

Jodi was thinking about all the things she has to be thankful for. She has a father and a mother who love her—

—a lot of friends to play with—

—and even a little brown dog named Fido.

JESUS
LOVES
ME

She has one friend who is very, very special. Each of us has this friend also. Do you know who it could be? It is Jesus. Jesus is our special friend and loves each one of us very much. He even died for us so that we can be resurrected and live again.

When Jesus died, He left us something special to help us remember Him. It is the Sacrament. Each Sunday we can go to Church and partake of the Sacrament. Each time we do this we are promising to remember Jesus and follow His Commandments.

Our Sunday School teachers have told us stories about Jesus and taught us His Commandments.

During the **Sacrament** we should sit very quietly in our chairs. We fold our arms and bow our heads and close our eyes while the Priests say the Sacrament prayers.

While the Deacons are passing the Sacrament to us we keep our chairs very still. We keep our mouths quiet, our arms folded, and our feet very, very still.

When it is our turn to take the bread and water, we take it with our right hand.

While we are sitting reverently, we can think about Jesus.

Jodi remembered that her teacher had told her many stories. She knew that Jesus loves little children and would call them to Him and bless them. He wants us to love one another and be kind and helpful.

Even when Jesus was a little child He was teaching the Gospel to others. One time his parents lost Him and when they found Him, He was in the temple teaching the men there.

Jodi decided she was going to be a good example and teach her friends to be more like Jesus.

Next Sunday, let's see if each of us can be like Jodi and get ready for the Sacrament, by—folding our arms,

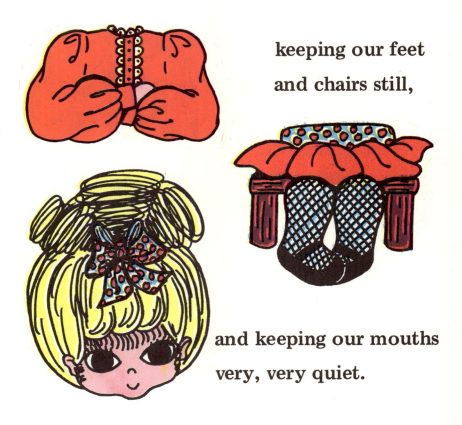

keeping our feet and chairs still,

and keeping our mouths very, very quiet.

Then we can think about Jesus and try to be good and live His Commandments each day.

50

Jeremy Learns To Sing

Jeremy was mad! When he got to Primary his teacher had asked him if he would sing in a special program in Sacrament Meeting.

Well, Jeremy didn't even like to sing!

In fact, whenever they had singing time in Primary he just sat and looked around.

Nobody was going to get him to sing! He couldn't see anything fun at all in singing a song.

On his way home from Primary that day, Jeremy heard a happy sound coming from a tree. He stopped and looked up to see what was happening.

What do you think he saw?

He saw the happiest little bird that he had ever seen. This little bird was sitting on a tree limb singing as loudly as could be. He was so happy he looked like he could just pop!

Jeremy looked at the little bird for awhile and thought, "He sure is having fun singing." Then he thought what a sad place the world would be if there weren't any little birds to sing.

By the time Jeremy got home he was glad the birds were singing, but <u>he</u> still wasn't going to sing in Church!

That night after Jeremy had said his prayers and hopped into bed, he could hear a noise outside his window.

He listened very closely and found out that it was some crickets singing in the bushes. Jeremy really liked to hear the way they sounded. They seemed so happy and friendly. In fact, their singing helped Jeremy fall asleep.

When Jeremy woke up the next morning, the first thing he heard was his dad singing in the shower. He always did that—every single morning! Jeremy decided his dad must like to sing—he sounded happy.

By now Jeremy could smell breakfast cooking, so he got dressed and hurried into the kitchen.

A few days later, Jeremy went back to Primary. During the singing time the Chorister told the children that singing is very important in Church. There are songs about Jesus, our families, the Gospel, the world, and many other things. She also said it makes Heavenly Father happy when we sing our praises to Him.

Well, Jeremy thought about how much happier the world was because of the birds and crickets singing, so he thought maybe he would sing just a tiny bit—not a lot—but maybe just once.

He sort of hid his face behind his shirt so nobody would know he was singing. He sang a few notes and looked around him.

He was really surprised to see that no one was paying any attention to him. In fact—they were all singing. Lisa looked so happy and was singing so loudly that Jeremy sat up straighter and sang out in his best voice.

Was he ever surprised!

He found out that singing is fun. It really made him feel happy inside.

He even told his Primary teacher that he would sing in the special program after all.

When Jeremy walked home, he felt
so good that he just had to stop at the
tree and sing with the birds. He had
learned that it's really fun to sing.